THE doors

THE doors

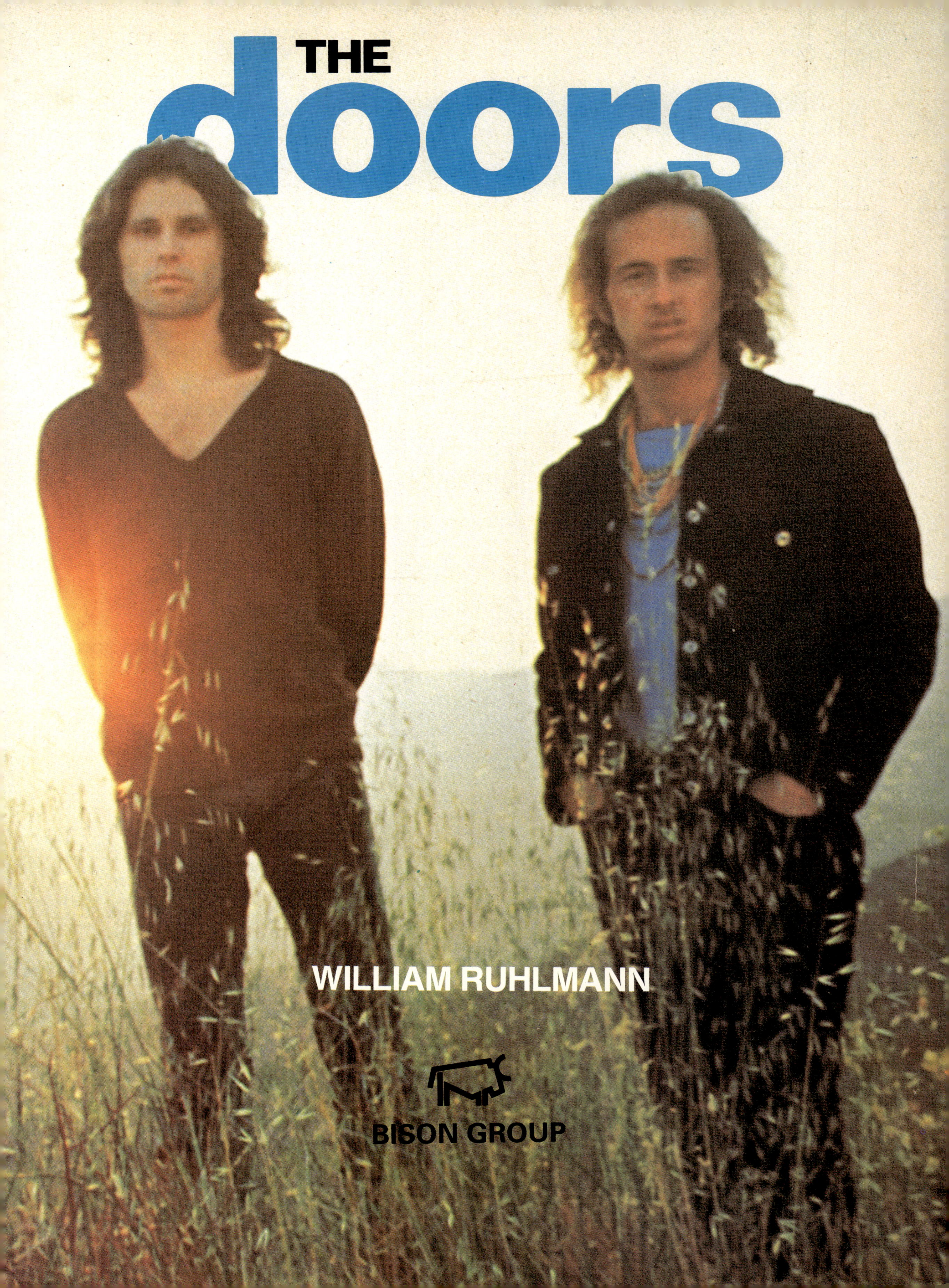

WILLIAM RUHLMANN

BISON GROUP

First published in 1991 by
Bison Books Ltd.
Kimbolton House
117A Fulham Road
London SW3 6RL

ISBN 0 86124 892 9

Printed in Hong Kong

Reprinted 1993

PICTURE CREDITS

AP/Wide World Photos: pages 9, 89.
The Bettmann Archive: page 15(top left).
Frank Bez/Globe Photo: page 65(right).
British Film Institute: page 59.
Brompton Photo Library: pages 1, 6(top right), 25, 27, 28, 29, 32, 33(bottom left), 34, 35, 36-37, 38-39, 40, 41(both), 48, 49, 56, 57, 58, 60-61, 63(right), 85, 90-91.
© Chris Fallo: pages 7, 26(top right), 46(bottom left, right), 70, 71(bottom left), 79, 82(both), 83, 84, 86, 87(top left).
Howard Frank: pages 13, 47, 50.
Steve Granitz/Retna Ltd: page 87(bottom).
Tony de Nonno/Globe Photo: page 81.
Jeffrey Mayer/Star File, Inc.: pages 17, 87(bottom).
Photofest: pages 26(bottom), 72-73, 76, 92(top left).
Ken Regan/Camera 5: pages 46(top), 71(top left, bottom right).
Joe Sia/Star File, Inc.: pages 24, 62(both), 63(top left, bottom left).
Springer/Bettmann Film Archive: page 6(bottom right).
Gloria Stavers/Star File, Inc.: pages 12(bottom), 16(bottom left), 33(top left).
Jay Thompson/Globe Photo: pages 74-75, 78.
UPI/Bettmann Newsphotos: pages 4-5, 8(both), 10-11, 15(top right, bottom), 16(top right, bottom right), 18, 19, 20-21, 22-23, 30-31, 43, 44(both), 45, 51, 52, 53, 66-67, 68, 69(top right), 80, 92(bottom).
U.S. Navy: page 12(top right).
University of California, Los Angeles; University Archives: page 14.
Vintage Magazine Company: page 42.
© Baron Wolman: pages 38(bottom left), 54(both), 55(both), 64, 93.

Contents

Page 1: Jim Morrison and Robby Krieger onstage during a performance in 1968.

Pages 2-3: Left to right: John Densmore, Ray Manzarek, Jim Morrison, and Robby Krieger in 1968.

Pages 4-5: The Doors captivated audiences for only a handful of years but, some twenty years after their last live and studio performances, their music continues to inspire musicians and rock fans everywhere.

Introduction

For four years, 1967-71, the Doors were among the most popular and critically acclaimed rock bands in the United States. In that period, they scored seven gold albums as well as the million-selling singles "Light My Fire," "Hello, I Love You," and "Touch Me." Just as fans loved their music, critics admired their originality and the poetic lyrics of lead singer Jim Morrison.

The Doors' ascension coincided with, and was affected by, the turbulent days of the late 1960s and early 1970s, with their anti-war protests, drug experimentation, and counter-cultural aspirations. The band's music reflected those times better than that of any of its contemporaries.

But the confrontational style of the era was also mirrored in the group's career, especially in the difficulties of its lead singer, the charismatic, unpredictable Morrison, who took the spirit of rebellion onstage with him and lived it offstage as well, finally with dire consequences. The Doors were as popular as ever when Morrison died suddenly in 1971.

The band itself broke up two years later, but regrouped in 1978 to record backing tracks to Morrison's poetry. When director Francis Coppola used the group's "The End" as a theme in his film *Apocalypse Now* in 1979, a full-scale Doors resurgence began, including a best-selling biography of Morrison, newly unearthed live audio and videotapes, massive media coverage, more gold and platinum records, and a 1991 film tracing the group's brilliant yet tempestuous career.

Now, 20 years after Morrison's death, with countless musicians exhibiting their influence, it's apparent that the Doors made not only the most timely music of the 1960s, but also the most timeless.

Above right: Doors lead singer Jim Morrison, circa 1970, at a point when he had begun to reject his teen idol image and adopt the serious countenance of a poet.

Right: Martin Sheen, as Captain Willard, in Francis Ford Coppola's film *Apocalypse Now*. Doors music on the film's soundtrack introduced the group to a new generation.

Below: During their four-year recording career, from January 1967 to April 1971, the Doors released eight record albums, six of which contained new, original studio material. Such an output was typical of the frenetic pace of record releases in the 1960s.

Below: Though the Doors achieved great popularity in Europe, they toured the continent only once, in summer 1968. On September 13 of that year, they taped a concert in front of the Frankfurt town hall for a West German TV show. This was at about the same time that their biggest European hit, "Hello, I Love You," was racing up the charts. They would return to Europe briefly for the Isle of Wight festival on August 30, 1970.

Right: Jim Morrison signs an autograph during a break in the taping of the Doors' performance on *4-3-2-1 Hot & Sweet*, a West German TV program, September 13, 1968.

Opposite: Morrison's grave at Père Lachaise cemetery in Paris, as it appeared in 1982, featuring a bust of the singer and extensive graffiti scrawled by fans, who reportedly visited the grave at the rate of 50-60 a day.

they have got the Guns But we have the NUMBER
THE UNKNOW SHOULDIER
1943-1971
LA WOMAN
This is the End
my only friend the End
JIM
POET

Poetry and Revolution: 1965-66

In *The Swing Era*, his massive study of the 1930-45 period of popular music in the United States, scholar and musician Gunther Schuller expresses amazement that a nation which could be so enamored of the "drivel" that made up the lyrics of 1930s and 1940s pop songs could also pull itself together sufficiently to overcome the Nazis in World War II. Though Schuller is no fan of rock 'n' roll, were he to examine the popular music of the second half of the 1960s he might conclude that, both in musical and lyrical terms, that music better expressed the crisis nature of its times than did the music of the swing era. Perhaps in the 1940s, those innocuous pop songs presented an escape to an audience for whom things were all too serious; but in the 1960s, audiences looked to the pop charts to find a mirror for the social

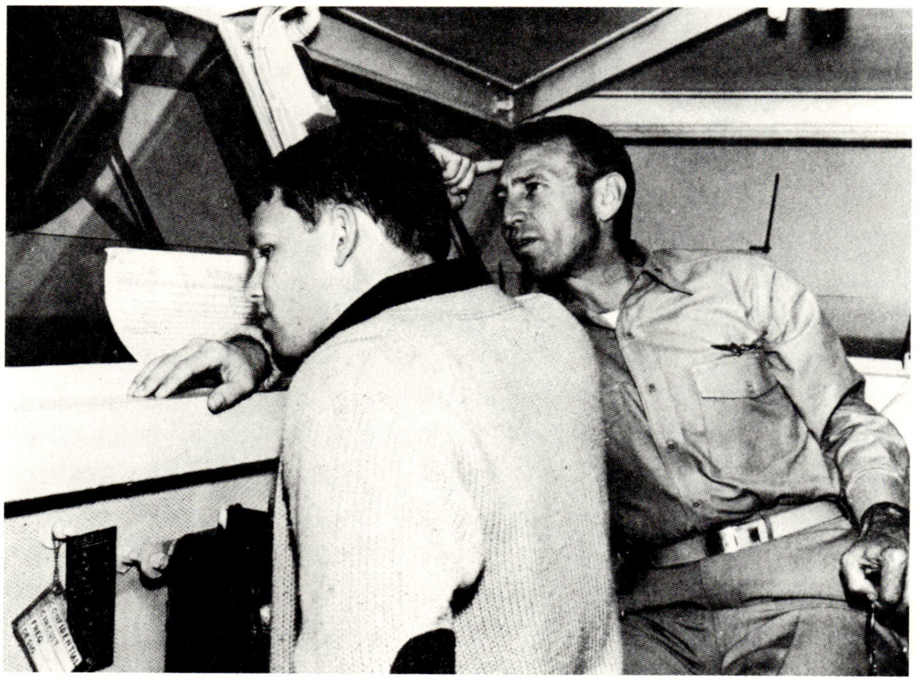

Above: Captain Steve Morrison, United States Navy, and his 20-year-old son Jim, on the bridge of the U.S.S. *Bon Homme Richard*, January 8, 1964. Prior to this photo being taken, the younger Morrison had been given a haircut by the ship's barber. It was probably the last time his hair would ever be so short.

Left: Jim Morrison three years later, in a photograph taken by *16* magazine editor Gloria Stavers, an expert if there ever was one on what young, female America wanted in its pop stars. It was seductive pictures such as this that, as much if not more than the music, helped make the Doors teen idols when they emerged in 1967.

Above: Though Morrison became the Doors' visual focus, the entire group had a sense of how its appearance contributed to its appeal, even if that appearance reflected the "anti-fashion" look of the times. Here, their unshaven, unkempt demeanor says they don't care whether you like them or not – but their autographs say they care very much.

turmoil that surrounded them. And though "drivel" remains a constant in some proportion of pop lyrics, even today, 25 years later, it isn't unusual for the best-selling records also to be the most thought-provoking.

This may suggest one reason why the music of the Doors – Jim Morrison, Ray Manzarek, John Densmore, and Robby Krieger – which was recorded from 1966 to 1971, not only had a profound impact on listeners when it was first heard, but has also proven perennially popular ever since. Though there are certainly other reasons why the Doors were a hit then and continue to be a hit today – among them lead singer Morrison's charisma, the group's uniquely spare, keyboard-dominated sound, and their many catchy songs – it is their seriousness of purpose, the artistic exploration in their music and lyrics, and their determination to confront the issues of their time (issues that, even in transmuted form, continue to affect us today) that cause them to continue to be important to listeners not old enough to remember when Morrison was still alive. In the parlance of the time, the Doors are still "relevant."

This is all the more surprising since, prior to 1965, when the band members first began to rehearse together in Los Angeles, it would have been hard to imagine music like theirs succeeding before a large audience. Even after winning World War II, most Americans made little connection between their social and political lives and the music they heard on the hit parade. And even after Elvis Presley arrived on the national scene in 1956, bringing with him a new look and a new beat, the *content* of pop music was very little changed. "Mairzy Doats" (a 1944 hit for five different performers) and Little Richard's 1956 hit "Tutti-Frutti" may be worlds apart in terms of musical style, but they're both nonsense songs.

Of course, not all popular music was devoid of lyrical significance. The Great Depression produced E.Y. Harburg and Jay

Gorney's "Brother, Can You Spare a Dime?", a Number One hit for both Bing Crosby and Rudy Vallee in 1932. The lyrics of Ira Gershwin, Cole Porter, and others often expressed a level of sophistication, even if they were romantically based, far beyond the usual moon-june doggerel of most pop. And Oscar Hammerstein II brought even more serious subjects to Broadway in shows such as *South Pacific* in the postwar era.

In another era, folk music, especially in the person of Woody Guthrie, spoke of important social issues in the lyrics of songs set to simple melodies. Though Guthrie himself (despite the influence he has since exerted) was not widely popular, his songs were sung by such national acts as the Weavers.

But it must be admitted that folk – at least topical folk – never became massively popular, and the social consciousness of Tin Pan Alley was never represented in more than a fraction of the songs produced. Even when the Beatles arrived in the U.S. in 1964, at first they were pursuing Presley's formula for success – the look and the music were different, but all they were singing about was teenage love.

Yet only a year later, the U.S. pop charts were awash with songs bearing much more serious, often political, sometimes poetic intents. The year's Top 10 hits included the Rolling Stones' "(I Can't Get No) Satisfaction," the Byrds' "Turn! Turn! Turn!" and "Mr. Tambourine Man," Barry McGuire's "Eve of Destruction," Bob Dylan's "Like a Rolling Stone" and "Positively 4th Street," and the Turtles' "It Ain't Me Babe."

The dominant influence here, of course, was Dylan and a style of music dubbed "folk-rock." Dylan had already made a major impression on the folk community in 1963 and 1964, and his music had reached the pop charts with Peter, Paul and Mary's version of "Blowin' in the Wind," but by 1965, Dylan had moved on from straightforward topical songwriting to a more ambitious, poetic style, and had added electric instruments to his sound. His lyrics, meanwhile, began to contain specific references to such writers as F. Scott Fitzgerald, Ezra Pound, and T.S. Eliot, and allusions to dozens more.

Above: The campus of the University of California at Los Angeles (UCLA) around the time in the mid-1960s that both Jim Morrison and Ray Manzarek attended the film school there.

Maybe none of this was unusual, given the times. If Dylan's music was folk-rock, his lyrics might be dubbed "college pop," though their author had dropped out of college himself. Nevertheless, he was an autodidact, and was writing for an educated audience. And clearly, more of an educated audience had come into existence. The middle-class baby boom generation of the immediate postwar era was now college age, and in a prosperous economy, college was affordable to the masses for the first time.

Influences on the Doors: Bob Dylan (in 1969, *above*) brought more serious subject matter and more poetic aspirations to pop-rock songs in the mid-1960s, helping lead to acceptance for Morrison's lyrics. Guru the Maharishi Mahesh Yogi (in 1967, *above right*) influenced Manzarek and Densmore, who studied with him in 1965. The Free Speech Movement (*right*), which erupted at Berkeley in December 1964, while Morrison and Manzarek were studying film at the L.A. campus, was the first of the student protests that marked an increasing politicization of young people, later reflected in the Doors' songs.

Then, too, as the civil rights struggle that had been gaining momentum since the mid-'50s began to grow increasingly urgent, and as the first stirrings of an opposition to the war in Vietnam (a war that directly affected young men in their late teens) began to be felt, topical issues seemed to penetrate the lives of Americans to an extent they had not since World War II.

Below: Doors producer Paul Rothchild once noted that he almost never saw Morrison without a book.

Right: One of Morrison's favorites was Beat author Jack Kerouac (shown in a 1960 photograph).

Page 18: Jim Morrison in an early publicity photo (taken in late 1966) for the Doors' debut album. This unassuming shot can be contrasted with photos taken only a few months later, which established Morrison as a rock god.

Page 19: In early performances, Jim Morrison often sang with his back to the audience. Once he turned around, he settled into his now-famous persona.

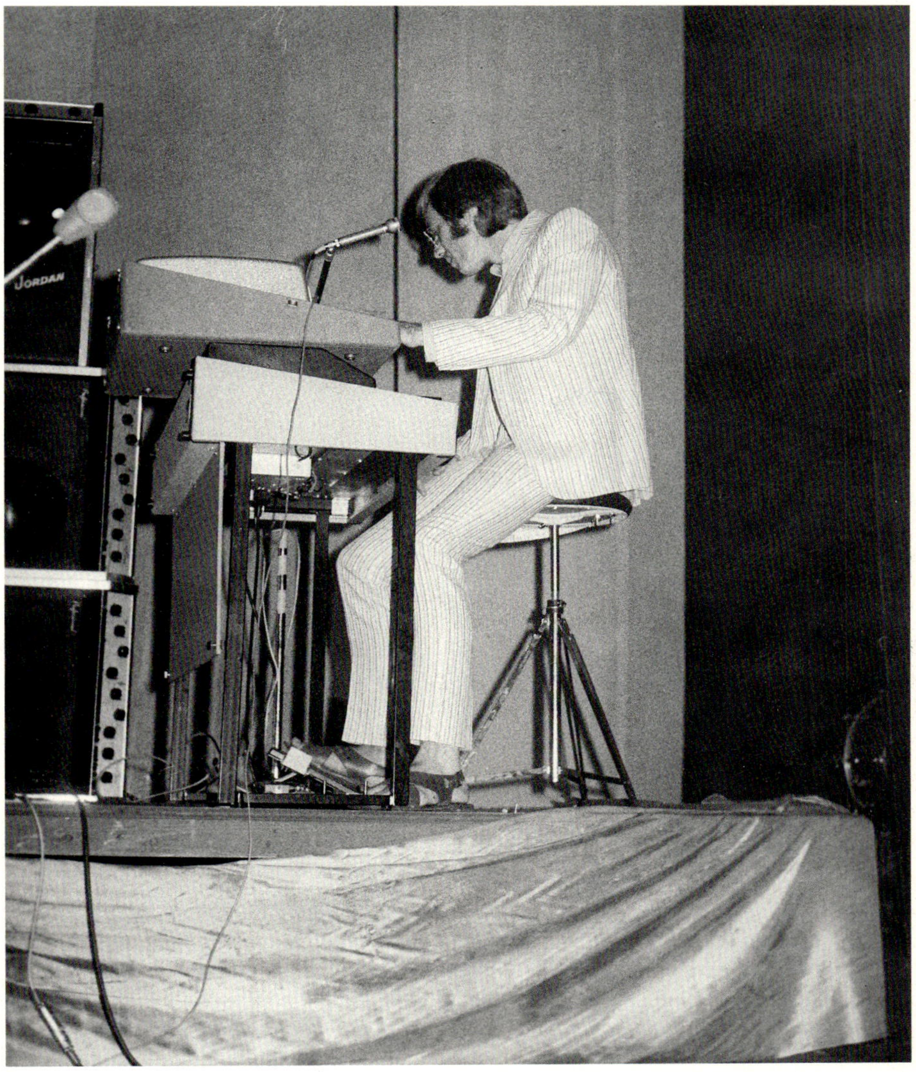

Above: Ray Manzarek, the oldest, best educated, and most musically adept of the Doors, organized the group out of his earlier band, Rick and the Ravens.

Opposite: British novelist and critic Aldous Huxley wrote the book *The Doors of Perception* that suggested the new group's name to Jim Morrison. Pity that, dying in 1963, Huxley never got the chance to hear them.

It is within this context that the 1965 formation of the Doors can best be understood. James Douglas Morrison, the son of a career naval officer who would in 1967 become the country's youngest admiral, was in many ways a typical young American of his time. In the summer of 1965, Morrison, who had been born December 8, 1943 and lived in various parts of the country, depending on where his father was stationed, had just graduated from UCLA film school, though his interests also extended to philosophy and poetry.

It was the latter, in which he exhibited the influence especially of the nineteenth-century French symbolists such as Rimbaud and Verlaine, that helped forge his musical relationship with Ray Manczarek, a fellow film school student and keyboard player. (Manczarek would simplify his name by dropping the "c" before the Doors achieved fame.) When Morrison recited his lyrics to "Moonlight Drive" to Manzarek, Manzarek made the sort of suggestion young men were making to each other all over the United States in the wake of the Beatles: He said they should start a band and make a million dollars.

Born in Chicago on February 12, 1935 (though he has been known to place the date in 1939 and even in 1942), Manzarek studied classical piano when he was growing up and graduated from DePaul University with a degree in economics. He had a stint in the army, and started at UCLA law school, but eventually ended up in the film school instead, meanwhile playing in a band called Rick and the Ravens with his two brothers. The Manczarek brothers were still in the band when Morrison became its lead singer and lyricist.

Manzarek added to the group a drummer and college student named John Densmore (born in Los Angeles on December 1, 1945) whom he had met at a Transcendental Meditation session with the Maharishi Mahesh Yogi (this was more than two years before the Beatles' infatuation with the Maharishi would make him a 1960s household name). Densmore, in turn, brought in guitarist Robby Krieger (born January 8, 1946), a friend from high school who had been in his previous band, the Psychedelic Rangers. With the departure of Manzarek's brothers and the group's bass player, the four members of the new group were in place.

Morrison had long since picked their name. "The Doors" was an immediate reference to *The Doors of Perception*, a short nonfiction book written by British novelist Aldous Huxley about his experiences with the drug mescaline, and in turn a reference to the poetry of William Blake. But it wasn't necessary to know the footnotes in order to appreciate the direct, one-syllable appeal of the name,

Left: Richard Burton and Tammy Grimes dancing at the Whisky A Go Go on October 26, 1965. At that time, playing the Whisky was the Doors' main career goal. They achieved it in the following year.

which was consistent with the then-current style in naming bands (the Byrds, the Who, etc.).

But if the group's name was in the style of the time, its music didn't seem to be. Having recorded a six-song demo while the Manczarek brothers were still in the band, the Doors had trouble placing it with a record label, and they also had trouble getting bookings in local clubs. One popular criticism was that they lacked a bass player, and it says something for their individuality that they never recruited one, instead having Manzarek play bass keyboard parts on a Fender Rhodes electric piano. (After their first album, however, their recordings usually did feature an added bassist, as the piano bass didn't record well.)

It wasn't until early 1966 that the group got its first club dates at the London Fog, in Los Angeles, where they were seen by Columbia Records A&R man Billy James and signed to a recording contract. But Columbia, which still hadn't made a firm commitment to rock despite their success with Paul Revere and the Raiders and the Byrds, let the Doors' six-month option lapse, by which time they had moved on to opening act slots at the more prestigious Whisky A Go Go. It was there they were seen by Jac Holzman, the head of Elektra Records.

Holzman, along with Maynard Solomon and Moses Asch, had been one of the main record label heads responsible for the folk boom of the early 1960s, releasing albums by such artists as Judy Collins and Phil Ochs. But by 1966, Holzman was looking to move into rock. Elektra had briefly signed the Lovin' Spoonful and the Byrds (under the name the Beefeaters), only to see them move on to commercial success with other companies. Holzman went to the Whisky A Go Go to see Love, a group led by Arthur Lee, and ended up signing the Doors as well, though he had to see them four times and hear them play a Krieger composition called "Light My Fire" before he was convinced. Holzman assigned Paul Rothchild, his staff producer and right-hand man, to handle the recording of the group's first album. As 1966 drew to a close, the Doors were on the verge of a success that would go beyond their hopes as well as their expectations.

Light My Fire: 1967-68

In early September 1966, the Doors and Elektra's Paul Rothchild repaired to Hollywood's Sunset Sound Recorders and spent two weeks recording the band's first album. As is true of most debut albums, the LP contained songs already worked up and arranged for the group's live set. As is not always the case, the record managed to capture the feel of live performing. In fact, the Doors' sound, which sometimes seemed a little thin live, here had a dramatic spareness, while their roots in jazz (especially Densmore) and classical music (especially Manzarek) came forward to meet the prevalent blues influence (especially Morrison) and straight rock (especially Krieger). The mixture was something genuinely new in music. From their first album, the Doors had a sound all their own, and this would prove to have positive and negative implications for them as their career went on.

Below: The Doors onstage in 1968. The staging and sound system are relatively primitive by today's standards, but with the focus on Morrison, the Doors' shows never lacked drama.

Opposite: A late 1967 Doors publicity photo, one of many with a bare-chested Morrison, emphasizing the group's sexual appeal.

Of course, a major element of that sound was Morrison's voice. Though Morrison has been celebrated as a frontman for his appearance and his theatricality, there has been a tendency to downplay the authoritativeness of his voice. At the time *The Doors* was released, Morrison was compared to the Animals' Eric Burdon, who also has a commanding baritone. But he was much closer to the blues singers Burdon had used as models, especially Howlin' Wolf (whose 1960 song "Back Door Man" was covered on the album). Just as Morrison had developed from an introverted stage performer who sang with his back to the audience, he had also developed a voice to go with his increasingly dramatic stage persona. It was a clear baritone with a precise articulation that could, unexpectedly, ascend to a blood-curdling scream and just as suddenly return to a distant, uninvolved, conversational tone.

Right: The Doors is one of three albums released in January 1967 to be awarded a gold record. The others: *Between The Buttons*, by the Rolling Stones, and *More Of The Monkees*.

Below: Despite his reputation for wild behavior, Jim Morrison was almost never photographed unless clean-shaven or with a full beard. Here's a rare look at the Morrison stubble.

Opposite: Proof positive of Morrison's infamous "shaman" abilities.

A part of the effect of that voice, naturally, was what it had to say, and it's hard to believe Morrison hadn't heard some of the Beat poets – Allen Ginsberg, Jack Kerouac (whose *On the Road* was an early Bible for him), Michael McClure (who would later become a friend) – not only because his poetic imagery often echoed theirs, but also because each of them had a sense of how to read expressively against spare, improvised music.

Morrison's lyrics (and he wrote most of them on the first album, omitting only Krieger's "Light My Fire," to which he added the "mire" and "pyre" lines) could have a sloganish brevity common to the best rock ["Break on Through (To the Other Side)"] and seemed rooted in the essentially romantic nature of much throwaway pop. But even when he was pleading for love, as so many crooners had before him, Morrison added an unusual poetic edge. It wasn't just that his way of putting it was to ask to "sleep all night in your soul kitchen"; it was his description of the world around him, of cars "stuffed with eyes," of a brain that seemed "bruised with numb surprise." The language had a distinct poetic rhythm and made use of such poetic devices as internal rhyme and alliteration, as in the couplet, "Your fingers weave quick minarets / Speak in secret alphabets." One might also note that Morrison's meanings were sometimes poetically obscure as well, but even in a pop world that had accustomed itself to the longwinded literary ambitions of Bob Dylan, his lyrics were something new and different.

Opposite: Jim Morrison onstage at the Hollywood Bowl, July 1968, a performance that was filmed and recorded, resulting in an album and music video released 19 years later.

Below: Jim Morrison and Robby Krieger, the Doors' two main songwriters, together onstage.

Pages 30-31: The first publicity photograph of the Doors to be issued by Elektra Records, upon the release of their debut album, *The Doors*, in January 1967.

As a singer, he was also deliberately daring. Pop had been accustomed to the expression of heavily implied sexuality since the subtle appeal of Frank Sinatra and the more overt smouldering of Elvis Presley. But Jim Morrison was not content to smoulder. The final chorus of "Light My Fire" had a practically animal savagery in distinct contrast to the classically derived Manzarek riff that followed it, and Morrison's infamous Oedipal lines in "The End" introduced a risqué subject (and, characteristically, an academic reference) new to pop music. In short, *The Doors* and the Doors represented something the popular music world had never heard before.

The critical reception accorded the group and the album said much about the state of journalism at the time. For, if the music had turned serious in the mid-1960s, a press had grown up that was ready to deal with it. Before, pop music was the province of teen magazines such as Gloria Stavers's *16*. Starting in 1966 and 1967, it became the province of Paul Williams's *Crawdaddy!* and others. The "straight" press, including *Time* and *Life* and the Los Angeles and New York *Times*, also began to pay more attention to pop.

The Doors were the ideal act to make use of such a transition. Their mixture of musical influences, along with the obvious poetic aspirations of Morrison's lyrics (and, as it turned out, his quick wit in press releases and interviews), endeared them to the new rock journalism. And as far as the still-powerful teen magazines were concerned, Jim Morrison was one cute guy.

The result, along with carefully selected concert appearances in L.A., San Francisco, and New York in the winter of 1966-67, gave the album a good send-off when it appeared in the first week of January 1967 – good enough that, even though the first single, "Break on Through (To the Other Side)" failed, the LP was able to hover in the low 100s in the charts after its chart debut on March 25.

Released in May, "Light My Fire" was in the charts by the end of the month and by August had hit Number One. The album, propelled by the single, reached Number Two. On September 11, both were certified gold (i.e., a million copies sold of "Light My Fire," $1 million worth of copies of *The Doors*).

Opposite: Perhaps the most famous photograph of Jim Morrison, taken in 1967, it was used on the cover of *The Best of the Doors* double album, released in 1987.

Above: *16*'s Gloria Stavers lent Morrison her fur coat for this shot, taken in New York in 1967.

Left: The Doors are presented with an award from *Billboard* magazine for the ascension of "Light My Fire" to Number One on July 29, 1967.

The success made the Doors the best-selling new rock group in a year that also introduced Buffalo Springfield, the Bee Gees, the Nitty Gritty Dirt Band, the Jimi Hendrix Experience, Traffic, Country Joe and the Fish, Big Brother and the Holding Company, Cream, the Grateful Dead, the Mothers of Invention, Pink Floyd, and the Who to the American charts. Many of these acts eventually went on to enormous success, of course, but in 1967 only the Doors (and Jefferson Airplane, who, though they had debuted in the charts in 1966, enlisted Grace Slick and found gold success with their second album, *Surrealistic Pillow* in 1967) managed that difficult twin victory of topping the teen-oriented singles charts and also being considered an "underground" band appropriate for listening by left-leaning, long-haired college students.

Pages 34 and 35: Two alternate shots from the 1967 photo session showing a sexy, shirtless Jim Morrison.

Pages 36 and 37: Though Morrison was Gloria Stavers' favorite photo subject of the period, here she provides a shot of the whole group in 1967, even if Morrison still stands out.

Below: Like the Doors, the Grateful Dead released its first album in 1967. It took four and half years to go gold. *The Doors* took only eight months.

Below: Country Joe and the Fish had a distinctive style, excellent musicians, and politically astute lyrics. The Doors had all that – and the charisma of Jim Morrison.

Left: Jimi Hendrix electrified rock fans in 1967, but never achieved the kind of pop singles success of the Doors.

Below: Janis Joplin's Big Brother and the Holding Company, also members of the class of '67, soon gave way to their singer's solo career (she is seen here with her Kosmic Blues Band).

Bottom: Of the '67 veterans, only the Jefferson Airplane duplicated the Doors' ability to top the pop singles charts and retain credibility with older, more serious rock fans.

It was, in a sense, a classic pop misunderstanding. Doubtless only a small percentage of the millions who tapped their feet to "Light My Fire" also heard the disturbing sentiments of "The End," and this dichotomy, between the band's mass pop image and its more complex, exploratory nature, would have great implications for the progress of its career. In a sense, though Krieger had brought in the hit single (and would compose much of the group's more commercial material) and Morrison represented the Doors' more esoteric side, the lead singer actually embodied both aspects. If the two sides were ultimately contradictory, they were no more contradictory than Morrison's alternately ambitious and reckless approach to his life and his work. Biographies of the Doors and Morrison have tended to concentrate on Morrison's legendary irresponsibility – his stupendous intake of drugs and drink, his onstage inconsistencies, and so forth – and alternately to excuse (on literary grounds) and judge his behavior. But without repeating the various stories, it will suffice to say that Morrison exhibited a conflicted attitude toward the fame he achieved in the Doors. And that attitude turned up as part of the Doors' music.

Contributing to the conflict was the era in which the Doors emerged. Nineteen sixty-seven may have been the year of the Summer of Love, but it was also a year of increasing death in Vietnam and increasing unrest about it at home. It was a year in which the racial tensions that had been rising in the U.S. for more than a decade boiled over into riots in such cities as Detroit and Newark (some said, to the strains of "Light My Fire"). Increasingly, the goal of peaceful integration espoused by Dr. Martin Luther King, Jr., was giving way to the militant, separatist, "burn, baby, burn" views of H. Rap Brown and others. In this context, Morrison's pithy description of the Doors as "erotic politicians," however trivial it may have been in reality, put the Doors on the extreme side of a polarized landscape. The group may never have joined a march or even

FAMILY DOG PRESENTS
FRI. ONLY $2.50
SEPT 29
LOTHAR AND THE HAND PEOPLE
$3.50 SAT. ONLY
SEPT 30
THE DOORS
1601 WEST EVANS ST.
DENVER COLORADO
CAPTAIN BEEFHEART
U.S. GRADE A
DANCE
& HIS MAGIC BAND
FRI & SAT
CONCERT
TICKET OUTLETS
DENVER: BRILLIG WORKS (2431 S. UNIVERSITY BLVD), POSTERS & THINGS (1215 -15TH, LARIMER SQUARE), HOBBIT BOOK SHOP, LTD (309 COLUMBINE, CHERRY CREEK), DENVER FOLKLORE CENTER (608 E. 17TH AV), BOB OWENS KUSTOM MUSIC CO. (6403 E. COLFAX), JERRY'S BOOK STORE (COLFAX & BROADWAY). BOULDER: CLANCY'S BOOK SHOP (1322 COLLEGE), PHANTASMOGORIA (1310 COLLEGE)
SCHNEPF '67
No 84-1
1967 © FAMILY DOG PRODUCTIONS 639 GOUGH ST., San Francisco, Calif. 94102

Opposite: Some bands stand the test of time; some do not. Here the Doors are co-billed with Lothar and the Hand People, September 29-30, 1967.

performed at a rally, but it seemed clear where they stood philosophically. And as the late 1960s wore on, that would be increasingly dangerous ground.

For the moment, however, the center held. Though *The Doors* would remain in the charts two years, and though it contained such likely single material as "Twentieth Century Fox," the music industry of the time demanded new product (not until the 1980s would several singles routinely be lifted from one album, as major acts took years between LP releases). The month after "Light My Fire" hit Number One, the Doors were back with a new single that prefaced the release of their second album: *Strange Days*.

The single was "People Are Strange."

Above: Somehow, at the height of the Summer of Love in 1967, the Doors, a group whose visual image suggested menace and whose songs spoke of patricide, were accepted by hippies along with their flower power favorites.

Though its lyrical theme of alienation made it an unusual choice for a pop single, it was undeniably catchy and, coming in the wake of "Light My Fire," was quickly added to radio playlists, even scraping into the Top 10. Starting with this single, the Doors' 45s tended to sell better than would be indicated by the Top 40 radio airplay they received. This can be discovered by comparing the records' chart positions in *Billboard*, which factors airplay into its ranking, with those in *Cash Box*, which is strictly a sales chart. In *Billboard*, "People Are Strange" peaked at Number 12; in *Cash Box*, at Number 10. In later years, Doors singles would routinely place 20 or more points higher in *Cash Box* than they would in *Billboard*.

THE HAYMARKET
COCKTAILS
RESTAURANT

Opposite: The Doors' music was as popular in Khe Sanh, South Vietnam as it was in San Francisco, California, its ominous, threatening tone accurately describing the lives of American GIs.

Opposite bottom: Chicago police beat demonstrators outside the Democratic Convention, August 28, 1968. The Doors' music was an equally effective soundtrack to the increasingly violent civil disobedience of the late 1960s.

Above: By 1969, the police had been superceded by the National Guard, who used bayonets to roust demonstrators during the "People's Park" demonstrations at Berkeley. "They've got the guns," sang Morrison, "but we've got the numbers."

Strange Days, which followed in October, bore many similarities to the debut album. Though the upgrading of Sunset Sound to eight tracks allowed for more elaborate production, the Doors' sound remained spare. Once again, the selections combined a melodic pop sound with Morrison's disturbing, sometimes opaque lyrics, touching frequently on death by drowning. The climactic song, taking up 11 of the LP's 35 minutes, was "When the Music's Over," a sequel to "The End" that strung together Morrison's poetic fragments with dramatic musical sections. *Strange Days* entered the charts in November, peaking at Number Three. A second single, Krieger's straightforward romantic tune "Love Me Two Times," became the group's third straight Top 25 hit, though it traced a downward trend in their chart peaks.

Bottom left: The Doors' second album, *Strange Days*, deliberately avoided putting a photograph of the group on the front cover, though there is a poster of the Doors on a wall.

Below: During performances of "The Unknown Soldier," Morrison would suddenly fall to the floor as if shot, though he was also known to sing from a prone position on occasion.

Bottom right: Though *Waiting For The Sun*, the Doors' third album, did not contain Morrison's "The Celebration Of The Lizard," an excerpt was included and the full lyrics printed on the album's inner sleeve.

Opposite: Jim Morrison in 1968. Still wearing the leather pants he would later abandon, Morrison nevertheless had begun to turn away from the beefcake look that had endeared the Doors to young fans.

Opposite: Jim Morrison onstage at the Fillmore East, rock's major East Coast concert hall, New York City, 1968.

Right: Jim Morrison in Los Angeles, circa early 1968. Even after the group's success, Morrison continued to live a somewhat vagabond, solitary existence, sleeping in seedy hotel rooms and on friends' couches.

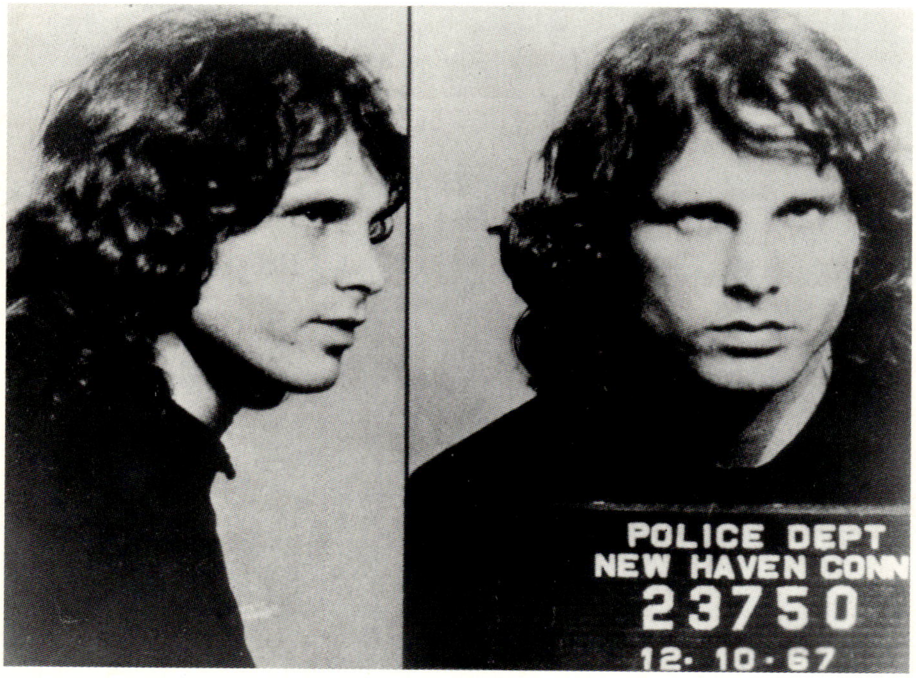

It may not have helped that, shortly after its release, Morrison was arrested onstage in a bizarre incident in New Haven, Connecticut. Having been sprayed with mace by a policeman who did not recognize him prior to the show, Morrison told the crowd about it and was arrested for breach of the peace, performing an indecent and immoral exhibition, and resisting arrest.

In retrospect, the New Haven arrest seems minor compared with the kind of trouble Morrison got into later, but it also stands as a signpost in the band's career. It would prove typical that the man who sang, "We want the world and we want it now" (in "When the Music's Over") would challenge authority in his life as well as his work. And it would also prove typical that he would use his often improvised onstage remarks to do so.

In the short run, the arrest probably didn't hurt the band's image, at least with the audience they were seeking. As 1968 approached, the polarization within the United States was reaching a peak and breaking out into battle. Shortly after the start of the year, the Tet Offensive in Vietnam changed the way many Americans thought

Above: Jim Morrison's mug shots, taken after he was arrested onstage in New Haven, Connecticut, a relatively minor incident compared to the troubles he would get into 15 months later in Miami, but nevertheless a harbinger for what life would be like for the Doors.

Right: The Doors conduct a press conference at London Airport where they arrived September 3, 1968 for a 17-day European concert tour.

Opposite: Ray Manzarek, Jim Morrison, and Robby Krieger in Frankfurt, September 13, 1968.

Above: Morrison's prodigious intake of intoxicants led to his missing one of the European concerts. The group went on as a trio, with Manzarek handling lead vocals.

about the war, helping to lead to the near-victory scored by peace candidate Senator Eugene McCarthy in the New Hampshire presidential primary.

The Doors' fifth single, released in March, was thus extremely well-timed, at least in the sense of current issues. Prefacing their third album, it was "The Unknown Soldier," a song that graphically depicted the soldier's death and that ended with Morrison declaring, "War is over." It wasn't, however, a sentiment the whole country wanted to hear, and the single barely made it into *Billboard*'s Top 40. (in *Cash Box* it got to Number 22.) That it was a hit at all is a testament to the Doors' ongoing popularity.

The single's involved production, with its sound effects and different sections, was a good indication of the work in progress for the third album. The pressure of producing a third record of original material within 18 months (unheard of today, but common to 1960s performers) caused a variety of problems. There were umpteen takes of songs. It was neces-

sary to interrupt the sessions to play concerts (the Doors never managed to put together a full-scale tour, but did go out regularly for a few days at a time). Densmore qu t the band, then returned. Morrison, frequently drunk, was dissatisfied with the band's decision to eliminate his next long composition, "The Celebration of the Lizard," from the record. (He too would briefly resign from the band during this period.)

It was what two band members would later call "the third album syndrome." Short on material, the Doors used some of their oldest unfinished songs, notably Morrison's "Hello, I Love You," which was released as their sixth single in June 1968. The album, *Waiting for the Sun*, followed in July.

Above and opposite: Jim Morrison in action at the Winterland Auditorium in San Francisco in December 1967. Morrison's inflammatory onstage remarks often caused near-riots at Doors concerts.

Despite the problems associated with the making of the album, its success was instantaneous. Both album and single went to Number One and were certified gold by the end of the summer. Mid-1968, in fact, marked the peak of the Doors' popularity. Hispanic singer Jose Feliciano even got to Number Three with a jazzy remake of "Light My Fire," while the Doors' version re-entered the charts.

At the same time, the group's concerts became more dangerous, sometimes resulting in near-riots. Though Morrison's provocative onstage behavior has been blamed for this, it must also be noted that the nature of rock concerts was in transition in this period, and some adjustment was probably inevitable.

The Doors were popular enough to play large auditoriums seating 10,000 or more, but the entertainment business had not yet adapted itself to such shows. True, the Beatles had toured to larger crowds years before, but their sound systems were woefully inadequate, and their audience of subteens was generally well-behaved when they weren't screaming.

Nineteen sixty-eight was a more confrontational time. If rock fans were rioting at Doors concerts, they were also facing off against police at anti-war rallies and, in August 1968, in the streets of Chicago during the Democratic National Convention. By the mid-1970s, despite occasional incidents, the rock touring business would be well-established, and arena owners would know what to expect and how to handle it. In 1968, a touring rock band had a far more difficult time.

None of which is to deny that Morrison was a provocative performer, often drunk, who teased audiences, and some of whose lyrics ("They've got the guns, but we've got the numbers" from the newly released "Five to One") played into the confrontational nature of the times. Morrison's interest in poetry and in film (a documentary of the band was now in the works) contrasted with his job as rock star. One can see in the extensive film footage of the Doors in performance during this period both his erratic and dramatic sense of stagecraft.

Opposite and below: Morrison's instability contrasted with the apparent dependability of the rest of the band, notably Manzarek (seen *below*), who often held things together, onstage and off.

Above and opposite: By the end of 1968, the Doors had escalated to the top of the charts and Morrison had become one of rock's most visible visionaries.

After several American concerts in July and August (among them a Fourth of July appearance at the Hollywood Bowl and a show at New York's Singer Bowl that were filmed and recorded), the Doors toured Europe in September and October, resulting in a BBC-TV film called *The Doors Are Open.* The film combined footage of the group performing at London's Roundhouse with footage of American demonstrators clashing with police. (The BBC wasn't the only one to make this connection: Filmmaker Jean-Luc Godard was just then combining footage of the Rolling Stones recording "Sympathy for the Devil" with political material for his film *One Plus One.*)

As the year drew to a close, the Doors edited their own documentary, played a variety of dates in the U.S. that only enhanced their reputation for disruptiveness, and released their first single in six months, "Touch Me." The single, a horn- and string-drenched song written by Krieger, would become their third major hit, reaching Number Three (Number One in *Cash Box*) and being certified gold early in the year.

The years 1967 and 1968 can be seen as a period of ascension for the Doors as America's most popular and most accomplished rock band. Like the early acts of a Shakespearean tragedy, these years were full of triumph as well as hints of trouble to come.

Touch Me: 1969-70

In a certain sense, the beginning of 1969 looked like the beginning of a new era after the increasing social difficult es of the previous few years in the United States. A new president, Richard Nixon, was inaugurated on January 20, and he had pledged both to bring the country together and to implement a "secret plan" to end the Vietnam War. For the Doors, coming off a year in which they had scored two Top Five albums and a Number One single, continued success seemed assured. They were working on their fourth album and planning their first full-scale concert tour. Of course, things didn't turn out as expected.

A year and a half before, the Living Theatre, led by Judith Malina and Julian Beck, had premiered a participatory play called *Paradise Now* in France. In February 1969, they brought the show to Los Angeles, where Jim Morrison sat in the front row for every performance. The play spoke of the need for a "Beautiful Non-Violent Anarchist Revolution," and featured performers who went down into the audience repeating phrases such as "I am not allowed to travel without a passport" and "I don't know how to stop the wars." In the text of the play, Phrase 5 is "I am not allowed to take my clothes off," and after it is spoken, the actors remove "as much of their clothing as the law allows."

By 1969, nudity had achieved the status of a revolutionary act. *Hair*, a musical that featured both nudity and actors dressed as policemen, had opened on Broadway the previous April and become a major hit. In November 1968, John Lennon and Yoko Ono had appeared nude on the cover of their album *Unfinished Music No.1 – Two Virgins*. Nevertheless, the Living Theatre was a highly controversial, avant-garde group, and it's fair to say that Morrison's version of their performance, enacted days later at the Dinner Key Auditorium in Miami, Florida, lacked the theatrical context in which *Paradise Now* was performed. Morrison had taunted audiences before, but his drunken rants on March 1,

Right, below right, and opposite: Jim Morrison in action during a concert in 1970.

Previous pages: Jim Morrison at Bronson Caves in the Hollywood Hills, 1970. This photograph was used as the cover of *American Prayer*, an album of Morrison poetry released in 1978.

Above: The cast of *Hair*, "the tribal love rock musical," which celebrated (some said, parodied) the counterculture and brought nudity to Broadway in 1968.

1969, represented what would be his most extreme performance.

According to writer Larry Mahoney, in the March 3, 1969, issue of *The Miami Herald,* "Morrison appeared to masturbate in full view of his audience, screamed obscenities and exposed himself." On March 5, a warrant was issued for the arrest of the singer, who had left Miami after the concert for a brief vacation in Jamaica. Although the charges had not been filed for four days after the event, Morrison was charged with unlawful

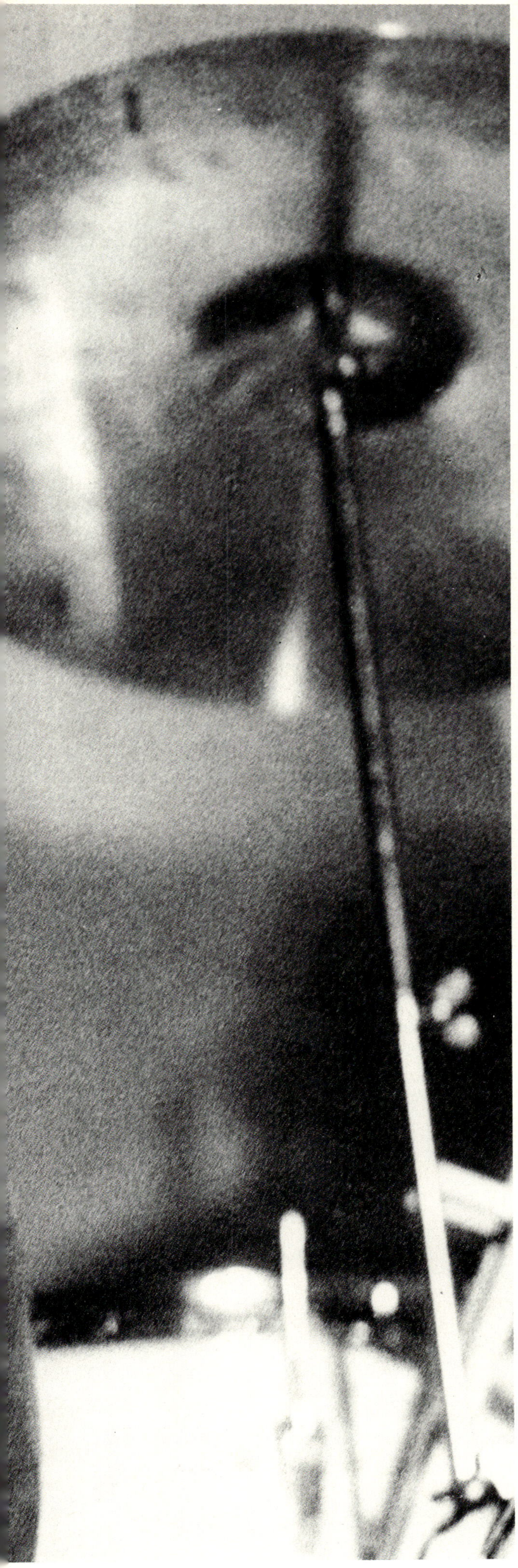

Left: A thoughtful Morrison waits for a rehearsal to begin. It was not unusual for Morrison to fail to appear at rehearsals, and sometimes even performances.

Above: Morrison took risks offstage as well as onstage, often hanging from hotel ledges by his fingers, and more than once falling to the ground.

flight to avoid prosecution and the matter was turned over to the FBI. When Morrison finally surrendered to authorities in Los Angeles on April 3, he was charged with "lewd and lascivious behavior, indecent exposure, open profanity, and public drunkenness."

The proposed tour was cancelled, and in future the Doors would be forced to post a bond to be forfeited in the event of illegal behavior at their shows. Police coverage would be so extensive that, in one instance, bandmembers were shown warrants for their arrest, already filled out and ready to be signed at the end of a show if necessary.

In the wake of the disaster, they attempted to carry on with their career. The next single, Krieger's "Wishful Sinful" (the first Doors record to list the actual author on the disc in

Pages 66 and 67: Jim Morrison and his attorney Max Fink arrive at the Los Angeles Federal Building April 14, 1969 for extradition proceedings that will return Morrison to Miami to face charges connected with the Doors' appearance there March 1.

Left: Demonstrating students did not go quietly, even after National Guard troops were called out to quell them. Here a Kent State demonstrator hurls a tear gas canister back toward Guardsmen.

Above: Nevertheless, having the guns proved to be a distinct advantage, as students learned May 4, 1970 at Kent State University, where four were shot dead during a demonstration.

stead of crediting it to "the Doors"), had been released in February. In May, it broke into the Top 30, but could rise no higher. Also in May, the third single from the still unreleased fourth album, "Tell All the People," another Krieger song, was released and only made it into the Top 40. The Doors' documentary *Feast of Friends* premiered in June, though, as a 16-millimeter, 40-minute film, it was unsuitable for widespread theatrical release. But the Doors returned to the album racks and the concert stages in July with the release of *The Soft Parade*, which was instantly certified gold, and an appearance July 25 at the Seattle Pop Festival. Though five of the album's nine songs had already been released, clearly revealing the eclectic intent of the record, and though the group had previously appeared onstage and on television with added instrumentation, *The Soft Parade* was widely criticized for its addition of horns, strings, and folk instruments to the Doors' sound. Those listening more closely might have noted an autobiographical tone to some of Morrison's lyrics, such as his asking for sanctuary and

Below: As if the Doors didn't have enough problems in 1969, their long-awaited fourth album, *The Soft Parade*, was widely criticized for departing from their familiar style. Fans bought it in droves anyway.

Bottom: When Elektra issued *Absolutely Live*, they altered the cover, superimposing a shot of Morrison in his leather pants from years earlier over the photo submitted by the band.

Below: The Doors' fifth album, *Morrison Hotel*, returned them to a spare, bluesy rock style and to critical favor, as they began to put their troubles behind them. The cover shot featured a real hotel in a seedy part of L.A.

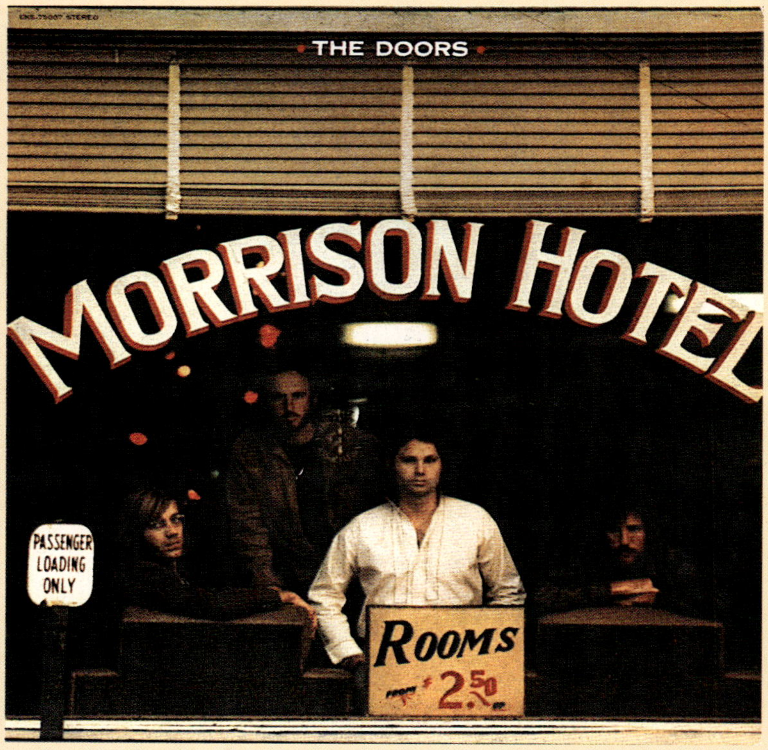

asylum in the title track, and his statement, "I can't take it anymore / The man is at the door."

The band's time and money were being spent on the troubles stemming from Miami, but they carried on, releasing the album's fourth single, "Runnin' Blue," in August (another low Top 40 entry) and appearing

This page: In the later days of the band, Morrison deliberately tried to downplay his sexual, photogenic image by adopting hats that obscured his face and repeatedly growing beards. In most cases, as on the Doors' greatest hits album, *13*, Elektra and the band prevailed on him to shave for an album cover photo shot.

at the Rock 'N' Roll Revival concert in Toronto in September that also marked the only appearance of John Lennon's Plastic Ono Band.

On November 11, during a flight to Phoenix, Arizona, to catch a Rolling Stones concert, Morrison and a friend were arrested and charged with interfering with the flight of an intercontinental aircraft (a newly enacted anti-skyjacking felony) and public drunkenness. By the following April, the charges would be dropped, but for the moment it was yet another legal albatross hung on the Doors' necks.

As 1969 ended, it had turned out to be the worst year in the band's history. Nor had the country fared that well either, with Nixon failing to end the war, protesters upgrading their protests to the monthly "moratoriums," and the Nixon Justice Department aggressively pursuing conspiracy charges against the

Pages 72-73: Morrison at Bronson Caves in the Hollywood Hills with his most frequent companion, Pamela Courson (who later adopted his name and called herself his wife). Courson died of a heroin overdose in 1974.

Chicago demonstrators and others. The much discussed "revolution" of the 1960s was turning bloody: The Students for a Democratic Society had mutated into the bomb-wielding Weathermen, and only four months hence National Guardsmen would be shooting students at Kent State.

As a new decade began, however, the Doors were still recording and doing concerts. On January 17 and 18, 1970, they appeared at New York's Felt Forum for shows that were taped for an upcoming live album. In February, they released their fifth album, *Morrison Hotel*, which was hailed as a hard-rock comeback and quickly became their fifth straight gold, Top 10 LP, though its single, "You Make Me Real," only got as high as Number 40.

In April, Simon & Schuster published Morrison's book of poetry, *The Lords and the New Creatures*, a combination of two privately printed volumes. The first contained brief prose poems, many of which talked about film and about the dichotomy between audiences and performers. "We are obsessed with heroes who live for us and whom we punish," Morrison wrote. The second section contained more imagistic poetry, much of it similar to Morrison's song lyrics. The book sold 15,000 copies in hardcover. (It has reportedly sold over 150,000 copies since its publication in paperback in 1971.)

In July, Elektra released *Absolutely Live*, a double album taken from concerts over the past year. Unlike most live albums, which are profit-making gestures featuring a performer's greatest hits plus applause, *Absolutely Live* was largely given over to previously unreleased material, including a version of "The Celebration of the Lizard." The group's hits were not included. Perhaps inevitably, Morrison's onstage comments included a reference to his alleged indecent act, when he told the audience they were in for a special treat. "No, no, no, not *that,*" he quickly added. "You only get that treat on full moons." *Abso-*

Opposite and below: In later performances, Morrison not only sported a full, unkempt beard, making him all but unrecognizable, but he also cut down drastically on his onstage movements, often hanging on the mike while declaiming his poetry.

lutely Live went gold the month of its release and reached the Top 10, though it stayed in the charts only 20 weeks.

On September 20, Jim Morrison was found guilty of indecent exposure and profanity in Miami, after the judge disallowed testimony regarding community standards. On October 29, he was found asleep on the porch of a woman he did not know in Los Angeles and was again charged with public drunkenness. On October 30, in Miami, he was sentenced to six months in jail and a fine of $500. The case was appealed.

In November, Elektra released a Doors greatest hits LP, *13*. Strangely unsuccessful for a Doors release, it got to only Number 25 and remained in the charts for only 21 weeks. It would take more than 18 months to go gold. Meanwhile, the Doors began work on a new album that would fulfill their seven-record deal with Elektra.

On his 27th birthday, December 8, 1970, Jim Morrison went into a studio and recorded a considerable amount of his poetry, using only a tambourine and instruments played by a couple of friends as accompaniment. Though nothing immediately came of this material, parts of it would be issued in the *American Prayer* album eight years later.

The Doors continued to play concerts sporadically. On December 12, they played what would be their last concert as a quartet in New Orleans.

For all intents and purposes, the end of 1970 brought the end of the Doors. Morrison, beginning with a desire for fame, had by now rejected his role as a rock star, and with the coming release of the new album, the Doors weren't even signed to a record label anymore.

But the coming year still held unexpected events for the group, and the coming decades even more surprise. The Doors may have ended as a group shortly after the end of 1970. But it wasn't long after they ceased being a contemporary phenomenon that they became a historic one.

The Legacy: 1971-1991

Below: Ray Manzarek has said that, at what turned out to be the Doors' last concert, Jim Morrison's spirit left him. Nevertheless, his body remained on earth another six months or so.

As far as Doors fans were concerned, 1971 began with the group at a peak of activity. In the previous 12 months, they had released three albums and toured extensively. If sales of the last two LPs had dropped slightly, it was only because one was an expensive double-record set, and the other a Christmas stocking stuffer of previously released material. A new album would set things right. If there were no immediate tour plans, it was only because the group was taking a welcome breather and because Jim Morrison's legal problems still had to be resolved.

Within the Doors camp, things didn't look so rosy, though it's hard to sort out the actual state of things. The band doesn't seem to have planned ahead at any point in their career, so their lack of any real plans in 1971 doesn't indicate much. Morrison left for France in March, apparently for a couple of months, while the three remaining members of the band finished mixing the new album and thereafter gathered to rehearse on a weekly basis.

John Densmore, for one, has suggested that the split between Morrison and the other three may have been permanent, no matter what else happened. At least, writing in his 1990 autobiography, *Riders on the Storm*, he says he hoped it was, that the Doors could go on without their lead singer.

But the public response to the new album suggested that the Doors' fans liked the quartet version of the band just fine. Released in April, *L.A. Woman* returned the Doors to Top 10 status. Its debut single, an edited version of "Love Her Madly," became their biggest hit in over two years. In June, Elektra released its followup, the moody "Riders on the Storm." It hit the charts the day Jim Morrison died.

Twenty years later, Morrison's death still has not been satisfactorily explained. At the time, the explanation was that he had died of a heart attack. The body was seen only by Morrison's companion, Pamela Courson,

Below: There was no indication to most fans in April, 1971 that *L.A. Woman*, which sounded like a comeback album, would prove to be the last recording made by the four Doors. The original album featured a cut-out on the cover, with the photo embossed on a piece of cellophane – an expensive packaging discarded on later printings.

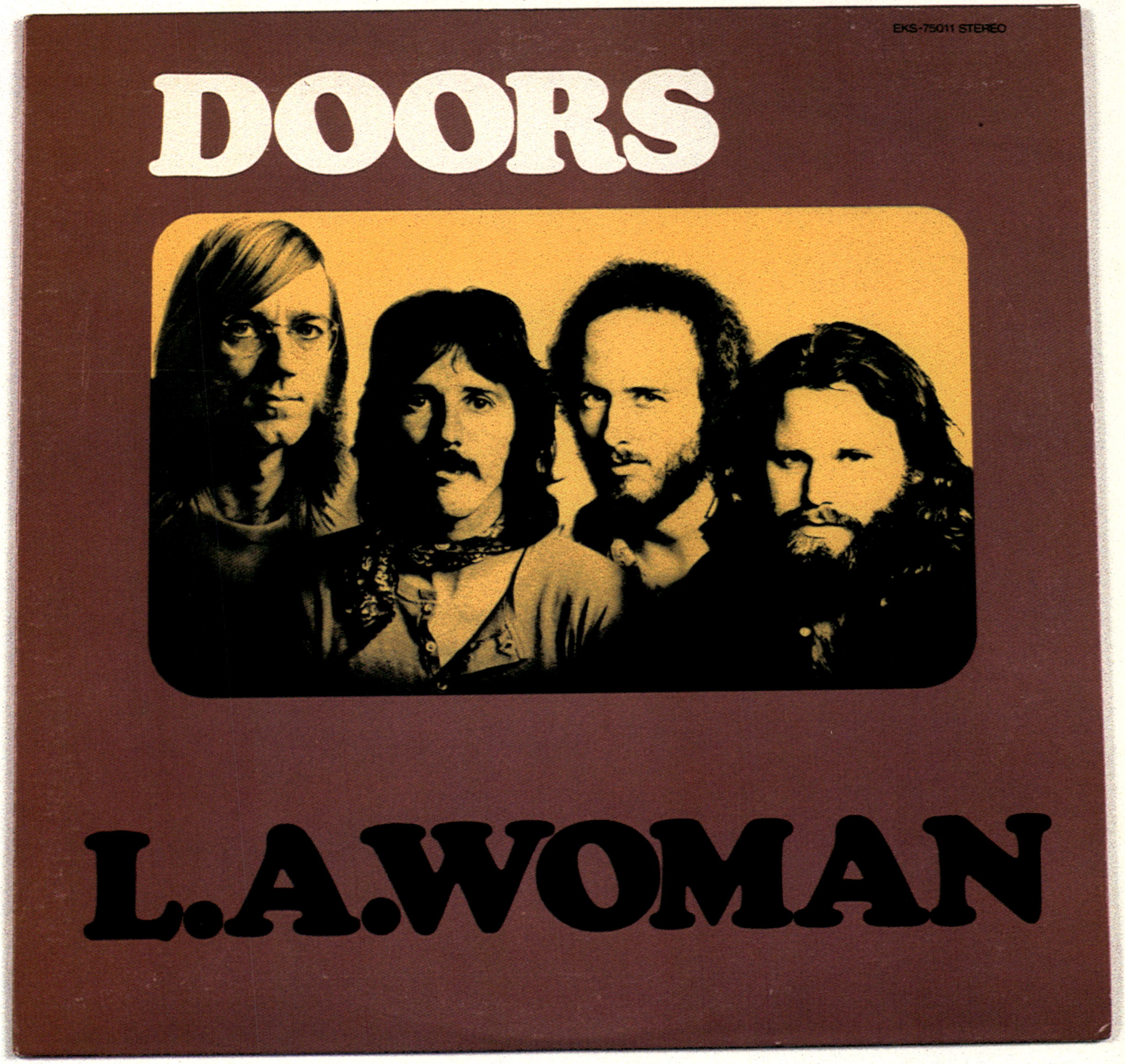

Opposite: The site of Morrison's July 7, 1971 burial, as it appeared two days later, when Morrison's death was officially announced to the press. Part of the mystery surrounding the death comes from the delay in announcing it – a full six days after it was said to have occurred.

Above: The Doors onstage after the death of Jim Morrison. Though the group recorded two albums and was warmly received in concert, they ultimately decided they needed a new frontman.

and the doctor who signed the death certificate. Speculation on whether Morrison died at all has continued to this day, to the extent that *No One Here Gets Out Alive*, the Morrison biography by Jerry Hopkins and Danny Sugerman published in 1980, even used the word "alleged" to describe it.

In recent years, however, circumstantial evidence has appeared that seems to settle the matter. One source for this, ironically, is Doors associate Sugerman who, in his 1989 memoir, *Wonderland Avenue*, recounts a conversation in which Courson first confesses to, then denies giving Morrison heroin on the night of his death. Another comes from Densmore's book. Densmore is incensed at the notion that Morrison might not be dead. He notes that Doors manager Bill Siddons, though he did not see Morrison's body in Paris when he flew over to make funeral arrangements, did discover a quantity of white powder that made him sick when he tried it. Densmore's conclusion is that the powder was heroin.

Below: The cover of the first post-Morrison Doors album, released only months after the singer's death, pointedly speaks of the group's determination to carry on, both in its title, *Other Voices*, and in the serious photograph of the band.

Below: It is no surprise that the Doors' renaissance has been fed by filmed accounts of the band's career and concert appearances. Jim Morrison lives on, on videotape.

Perhaps most tellingly, Densmore recounts an interview he did in 1986 with Roger Stephens in which Stephens told Densmore that he had been told by singer Marianne Faithfull and a French count named Jean DeBretti who "was one of Pamela's lovers" that they had been called by Courson to help break down the bathroom door and had discovered Morrison dead. Neither Faithfull nor DeBretti has yet confirmed this story.

Dead or not, Morrison has not been heard from since, and a sealed casket said to contain his mortal remains was buried at the Lachaise cemetery in Paris on July 7, 1971. That the Doors had intended to go on as a trio in any case seems confirmed by their subsequent actions. They rapidly announced they would go on, and re-signed to Elektra. Their first album, pointedly entitled *Other Voices*, was released less than four months after Morrison's death, and the Doors toured in the fall of 1971 to appreciative audiences. The LP's reviews stated the obvious – that Manzarek and Krieger didn't measure up to Morrison as singers and that their lyrics didn't, either. Nevertheless, the album got to Number 31 in *Billboard* and, for the moment, the Doors carried on.

Above: Compilation albums on LP and CD in the 1970s and 1980s have repackaged familiar Doors material successfully. Note that the covers always feature pictures of Morrison rather than the group as a whole.

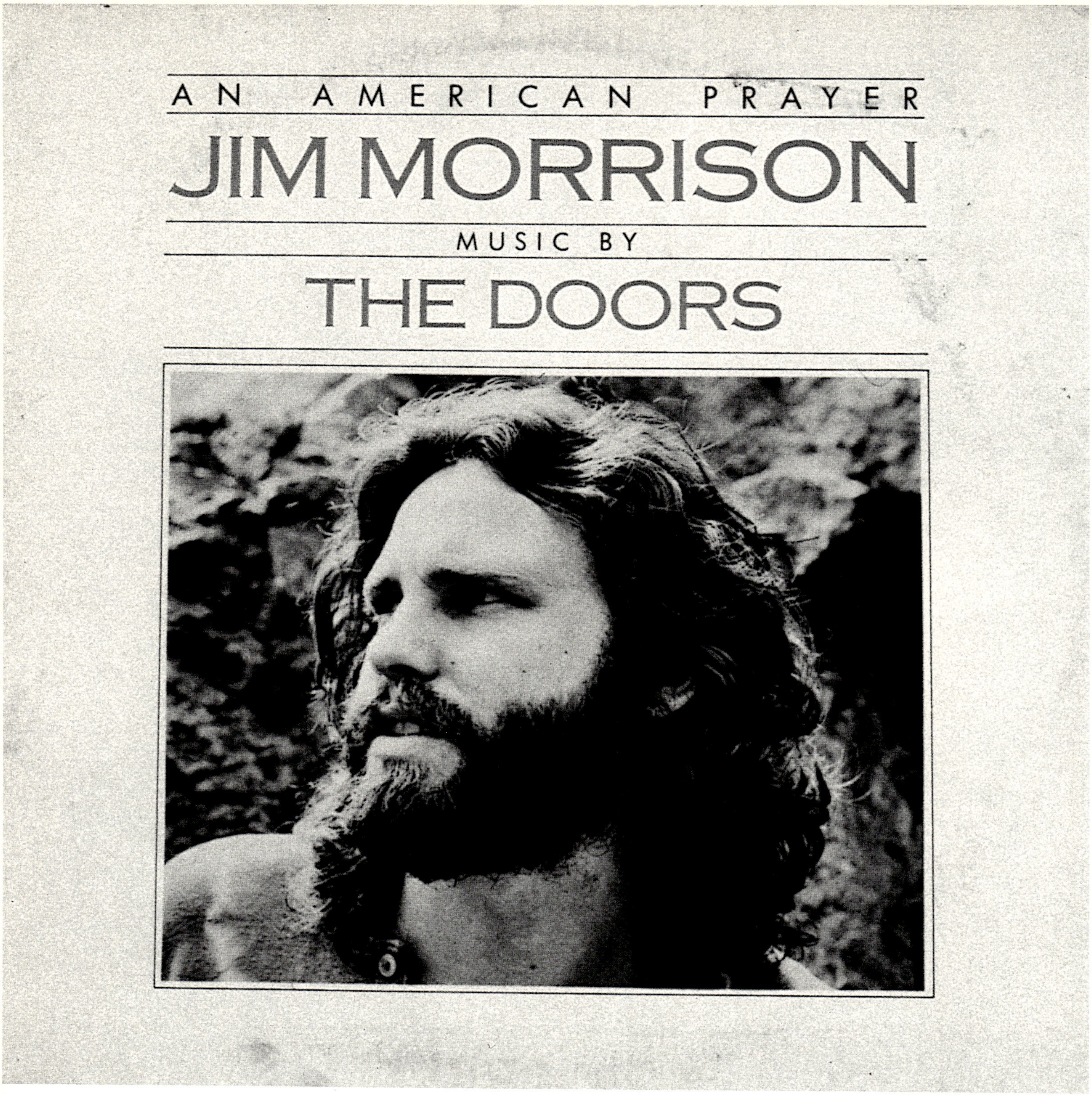

Elektra didn't help their case by releasing a second Doors compilation, *Weird Scenes Inside the Gold Mine*, with a cover illustration of Morrison, in January 1972. The two-record set marked the first LP appearance of two Doors B-sides, but otherwise gathered previously released album cuts in a new sequence. It only reached Number 55.

The Doors toured Europe in the spring of 1972, and in July released their second post-Morrison album, *Full Circle*. When it proved less of a success than the first one, they went to England in search of a new lead singer at the end of 1972. Manzarek then returned to Los Angeles, splitting with Densmore and Krieger.

Above: The first previously unreleased Morrison performances in seven years appeared in 1978 with the release of *American Prayer*, in which Morrison's poetry recitations were backed by newly recorded instrumental tracks by the three surviving Doors. The album was surprisingly successful.

The drummer and guitarist enlisted singer/guitarist Jess Roden, bassist Phil Chen, and keyboard player Roy Davies and renamed themselves the Butts Band. Their self-titled debut album appeared on Blue Thumb Records and failed to chart.

Elektra, meanwhile, responding to the audio fad for "quadrophonic" (i.e., four-speaker) sound, released a quad Doors compilation, *The Best of the Doors*. Its meager

Below: Martin Sheen in *Apocalypse Now*, which made excellent use of the sometimes sinister sound of the Doors' music. Film director Francis Ford Coppola attended UCLA film school at the same time as Jim Morrison.

sales suggested that, more than two years after Morrison's death, the Doors had ceased to matter to record buyers.

Manzarek, meanwhile, was preparing for a solo career, and he started it with *The Golden Scarab*, a concept album about Egyptian deities that featured him in gold facial makeup on the cover. It failed to sell, though its followup late in 1974, *The Whole Thing Started with Rock & Roll Now It's Out of Control* got to Number 150 in early 1975. (The album featured a reading of Jim Morrison's poetry by rock poet Patti Smith, who would make her own mark later in the year.)

In early 1975, Densmore and Krieger reemerged with a new edition of the Butts Band and released *Hear & Now!*, which disappeared without a trace. The band split, and Densmore began taking acting classes. Krieger stayed in the music business, but his next effort, *Robbie Krieger and Friends*, released in early 1977, was a jazz-rock fusion album of instrumentals. Manzarek, meanwhile, formed a new band, Nite City, which featured future Blondie bassist Nigel Harrison, and released *Nite City*. Neither effort

Below: Former members of the Doors have recorded a variety of solo and group works, especially during the 1970s. None has matched the artistic or commercial success of the Doors.

Above and right: In recent years, Elektra and the former Doors have managed to unearth some previously unreleased live recordings and have issued them on albums (*above*) to help meet the demand for Doors music. That demand also extends to accounts of the band's history, and drummer John Densmore (*right*) published his memoir on the band, *Riders On The Storm*, in 1990.

reached the charts. (A second Nite City album was released only in West Germany in 1978.)

By mid-1977, then, with the pop world given over to heavy metal, punk/new wave, and disco, the Doors were at a low point. The Morrison records weren't selling anymore, and the group's individual musical careers hadn't caught fire. A new Doors album didn't seem much of a commercial proposition, but that's exactly what the group began talking about. The idea was to resurrect the poetry tapes from Morrison's 27th birthday and set them to music. The project took longer than expected, but in November 1978 Elektra Records released *An American Prayer*, credited to "Jim Morrison, music by the Doors."

The album was not well-received critically, but sold surprisingly well, reaching Number 54. Still, if there was anything that helped spark a new Doors renaissance, it was film director Francis Coppola's use of "The End" in his Vietnam epic *Apocalypse Now*, which opened in selected theaters on August 15, 1979. The Doors' music had been played extensively by GIs in Vietnam, and in the images of napalm destruction Coppola used, the band's dark, ominous sound found a perfect visual equivalent. As the movie reached a broad audience during the next year, a new generation that had never heard the Doors' music wanted to know who performed "The End."

As a result, the Doors' first album, which contained "The End," re-entered the *Billboard* charts. *Weird Scenes Inside the Gold Mine*, also containing the song, went gold, more than eight years after its release, on September 15, 1980. The same month, Elektra released *Greatest Hits*, its fourth Doors compilation. The album soared into the Top 20 and went gold by the end of the year. (It would go platinum in 1981.) Nineteen eighty also marked the publication of the Hopkins/Sugerman biography of Morrison, and it topped bestseller charts.

Below: Most active on the musical front of the former Doors, guitarist Robby Krieger continues to record and to tour, with his son Waylon singing a few Doors songs in the show.

The surviving Doors, of course, helped to fan the flames, leading a graveside tribute to Morrison in Paris on the 10th anniversary of his death, as national magazines featured photos of the singer and articles on the phenomenon. The group also went in search of previously unreleased material, resulting in the October 1983 release of *Alive, She Cried*, the Doors' second live album, which reached Number 23.

The following years saw more books, videos, and albums, while the old albums continued to sell. When Elektra submitted the Doors' catalog to the Record Industry Association of America for new gold and platinum certifications in 1987, the results were dramatic: The 1980 *Greatest Hits* album, *L.A. Woman*, and *The Doors* were all certified for sales over two million; *Waiting for the Sun*, *The Soft Parade, 13*, and *The Best of the Doors* had all sold over a million copies; and *Alive, She Cried* was gold.

Below: Ray Manzarek (*left*) continues to involve himself in producing rock bands (he handled the 'boards for X, among others, in the 1980s), while John Densmore has turned his attention to acting in recent years.

These pages: Directed by Oliver Stone and starring Val Kilmer as Morrison, *The Doors* was released in the U.S. on March 1, 1991. Twenty years after Morrison's death, the continuing popularity of the Doors made the band the subject of a major motion picture.

Elektra promptly released two new Doors albums, *Live at the Hollywood Bowl* (with an accompanying video) and a revised version of *The Best of the Doors* expanded to double CD length. The group's albums had begun to be released on CD in the mid-1980s with, as was typical of major-label CD reissues of the time, disappointing results. The catalog was remastered and came out on CD in 1989.

Today, 20 years after Jim Morrison's death, the Doors' reputation continues to flourish. Previously unpublished Morrison writings have been issued in two posthumous books, *Wilderness* (1988) and *The American Night* (1990). Densmore has published his autobiography, and Manzarek and Krieger have occupied themselves producing other artists and issuing the occasional solo release. Nineteen ninety-one sees the release of Oliver Stone's film biography *The Doors* starring Val Kilmer as Morrison, more books about Morrison and the group, and a boxed set of Doors recordings from Elektra.

Left: Pre-punk Patti Smith, like Morrison, combined a desire to be a poet with a desire to be a rock singer, and like him, achieved both. She also recited Morrison's poetry on a Ray Manzarek album.

It is probably fair to say that a large part of the band's continuing appeal rests on the eternal irresponsible-adolescent appeal of Jim Morrison – his good looks and his reputation for dissipation. But just as much of it rests on the Doors' music, which, for a band so finely attuned to the social and musical trends of its times, is amazing. Yet the Doors today sound anything but dated. Timely they may have been, but by now they have also proven to be timeless.

Below: Not only the Doors' music itself became popular in the 1980s – rising groups also borrowed some of their style. U2's lead singer even took to evoking Morrison's name onstage.

Opposite: Jim Morrison remains a controversial figure, yet no one can deny the tremendous effect that his work with the Doors has had on rock music.

Discography

Record Label	*Record Number*	*Title*	*Release Date*
Singles			
Elektra	E-45611	Break on Through/End of the Night	1/67
Elektra	E-45615	Light My Fire/The Crystal Ship	5/67
Elektra	E-45621	People Are Strange/Unhappy Girl	9/67
Elektra	E-45624	Love Me Two Times/Moonlight Drive	11/67
Elektra	E-45628	The Unknown Soldier/We Could Be So Good Together	3/68
Elektra	E-45635	Hello, I Love You/Love Street	6/68
Elektra	E-45646	Touch Me/Wild Child	12/68
Elektra	E-45656	Wishful Sinful/Who Scared You	2/69
Elektra	E-45663	Tell All the People/Easy Ride	5/69
Elektra	E-45675	Runnin' Blue/Do It	8/69
Elektra	E-45685	You Make Me Real/Roadhouse Blues	3/70
Elektra	E-45726	Love Her Madly/(You Need Meat) Don't Go No Further	3/71
Elektra	E-45051	Light My Fire/Love Me Two Times	4/71
Elektra	E-45052	Touch Me/Hello, I Love You	4/71
Elektra	E-45738	Riders on the Storm/Changeling	6/71
Elektra	E-45757	Tightrope Ride/Variety Is the Spice of Life	11/71
Elektra	E-45807	The Mosquito/It Slipped My Mind	8/72
Elektra	E-45059	Riders on the Storm/Love Her Madly	9/72
Elektra	E-46005	Roadhouse Blues/Albinoni Adagio	1/79
Elektra	69770	Gloria/Love Me Two Times	11/83

Record Label	*Record Number*	*Title*	*Release Date*
Doors-Related Singles			
Rhino	RNTI 403	Kinky Reggae/Get Up Stand Up (by the Krieger-Densmore Reggae Bonanza)	1983

Record Label	Record Number	Title	Release Date
Albums			
Elektra	EKS-74007	The Doors	1/67
Elektra	EKS-74014	Strange Days	10/67
Elektra	EKS-74024	Waiting for the Sun	7/68
Elektra	EKS-75005	The Soft Parade	7/69
Elektra	EKS-75007	Morrison Hotel	2/70
Elektra	EKS-2-9002	Absolutely Live	7/70
Elektra	EKS-74079	13	11/70
Elektra	EKS-75011	L.A. Woman	4/71
Elektra	EKS-75017	Other Voices	11/71
Elektra	EKS-2-6001	Weird Scenes Inside the Gold Mine	1/72
Elektra	EKS-75038	Full Circle	7/72
Elektra	EQ-5035	The Best of the Doors	9/73
Elektra	5E-515	The Doors' Greatest Hits	9/80
Elektra	60269	Alive, She Cried	10/83
Elektra	60417	Classics	7/85
Elektra	60741	Live at the Hollywood Bowl	6/87
Elektra	60345	The Best of the Doors	7/87

Record Label	Record Number	Title	Release Date
Doors-Related Albums			
Blue Thumb	BTS-63	The Butts Band (featuring Robby Krieger and John Densmore)	1974
Mercury	SRM-1-703	The Golden Scarab (by Ray Manzarek)	1974
Mercury	SRM-1-1014	The Whole Thing Started with Rock & Roll, Now It's Out of Control (by Ray Manzarek)	1974
Blue Thumb	BTS-6018	Hear & Now! (by the Butts Band, featuring Robby Krieger and John Densmore)	1975
Blue Note	LA774	Robbie Krieger and Friends (also features John Densmore)	1977
20th Century-Fox	T-528	Nite City (featuring Ray Manzarek)	1977
20th Century-Fox	6370 263	Golden Days & Diamond Nights (by Nite City, featuring Ray Manzarek) (released only in West Germany)	1978
Elektra	5E-502	An American Prayer (by Jim Morrison, music by the Doors)	11/78
Passport	PB6017	Versions (by Robby Krieger, featuring Ray Manzarek and John Densmore)	1983
A&M	SP-4945	Carmina Burana (by Ray Manzarek)	1983
Cafe	730	Robby Krieger	1985
I.R.S.	82004	No Habla (by Robby Krieger)	1989
I.R.S.	82014	Door Jams (by Robby Krieger)	1989

Index

Page numbers in *italics* indicate illustrations